BALLOON SKEWER

Written by
Read With You Science Academy

Read With You

Published by Read With You Publishing
Designed by Read With You Science Academy

ISBN-13: 978-1-949391-75-6
ISBN-10: 1-949391-75-2

Second Edition April 2021
Printed in the United States of America.

Welcome to the

READ WITH YOU SCIENCE ACADEMY!

Today we will do an experiment called

BALLOON SKEWER

Before we begin, let's review some important words:

skewer:
A thin wood stick with a sharp point.

to pierce:
To make a hole with something sharp, like a skewer.

hypothesis:
What we think will happen in the experiment; a guess.

Remember:
Be a careful scientist! Ask an adult to help you with the experiment.

QUESTION:

What I want to know.

Will the balloon pop when I pierce it with a skewer?

- [ ] Yes, the balloon will pop when I pierce it with a skewer.
- [ ] No, the balloon will not pop when I pierce it with a skewer.

HYPOTHESIS:

What I think will happen and why.

I think the balloon **will pop** because...

I think the balloon **will not pop** because...

The balloon **will** pop:	The balloon **will not** pop:
The balloon will pop because the skewer is very sharp.	The balloon will not pop because the skewer can't get through the balloon.
The balloon will pop because the air wants to come out.	The balloon will not pop because the skewer stops air from leaving the balloon.

What do you think?
Circle your hypothesis!

EXPERIMENT:

How we will test our hypothesis.

THINGS WE NEED

- 1 permanent marker
- 3 skewers
- At least three 12-inch latex balloons
- A bowl with some water
- Liquid dish soap

permanent marker

skewers

balloons

bowl

water

dish soap

PROCEDURE:
The steps to complete the experiment.

Step 1A

- First, **inflate** the balloon until it is halfway filled with air.
- Tie a knot at the end of the balloon.
- Remember to ask an adult for help if you need it!

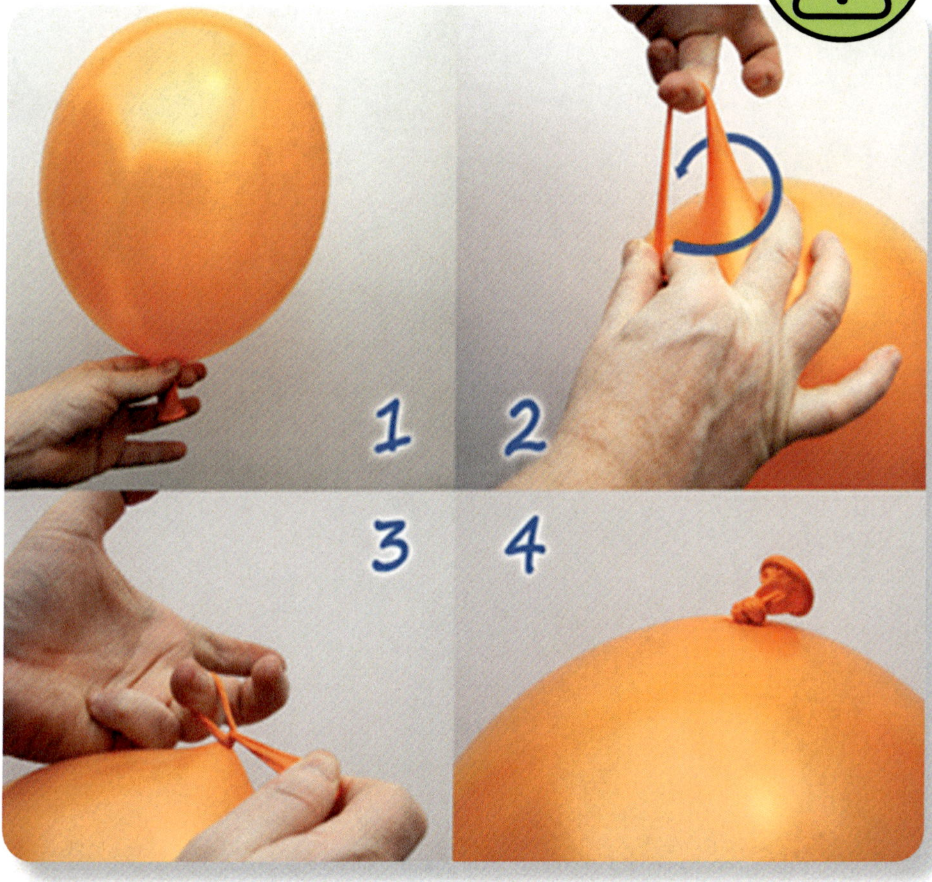

Step 2A

- **Examine** the balloon. Hold the balloon so the knot you tied is at the bottom. Now, look at the top and bottom of the balloon. Do you see the area at both ends where the rubber is thicker? This is where you will pierce the balloon with the skewer. But, don't pierce it yet!

Step 3A

- Add 3 or 4 drops of dish soap to the water in the bowl and mix well. Dip a skewer into the soapy water. This will help the skewer to go through the balloon. The whole skewer should be covered with the soapy water.

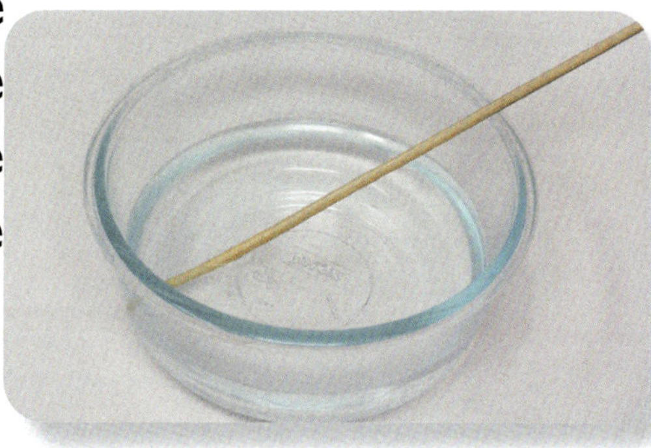

Step 4A

- Find the place where the rubber is thicker on the bottom of the balloon. Place the pointed end of the skewer here, and push it into the balloon. Use gentle pressure and a little twisting motion to pierce the balloon. Remember to be a careful scientist and ask for help from an adult if you need it! Be careful not to jab the balloon or yourself.

Step 5A

- Push the skewer through the balloon until the pointed tip of the skewer is touching the top of the balloon. Then, find the other place where the rubber is thicker at the top of the balloon. Keep pushing until the skewer pierces the rubber. Ta-da! You did it!

Step 6A

- Gently remove the skewer from the balloon. The air will leak out of the balloon, but the balloon will not pop.

Write down what you see.

In your science journal,

write your results!

- What was your hypothesis? (Look on page 5 of this book!)
- Was your hypothesis correct?
- Where did you pierce the balloon?
- How did you do it?
- Did the balloon pop?

Let's do it again!

Did you notice there are parts of the balloon where the rubber is not thick like at the top and bottom? Where do you think is the best place to pierce the balloon if you don't want it to pop? Let's do an experiment to find out.

QUESTION:

What I want to know.

Where is the best place to pierce the balloon if I don't want it to pop?

HYPOTHESIS:

What I think will happen and why.

I **think** the top and bottom of the balloon are the best places to pierce it because...

I **think** the sides of the balloon are the best places to pierce it because...

What do you think?

Write your new hypothesis in your science journal.

PROCEDURE:

The steps to complete the experiment.

Step 1B

Before inflating the balloon, use the permanent marker to draw some dots on the balloon. Draw dots at the top, bottom, and middle of the balloon. They should all be the same size. This will help you see the hidden "stress" in a balloon.

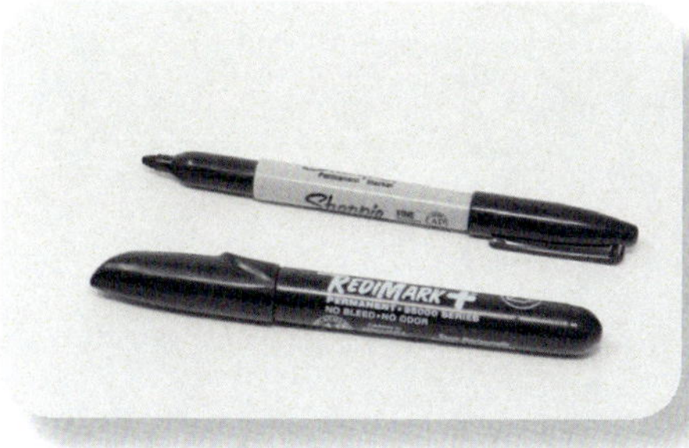

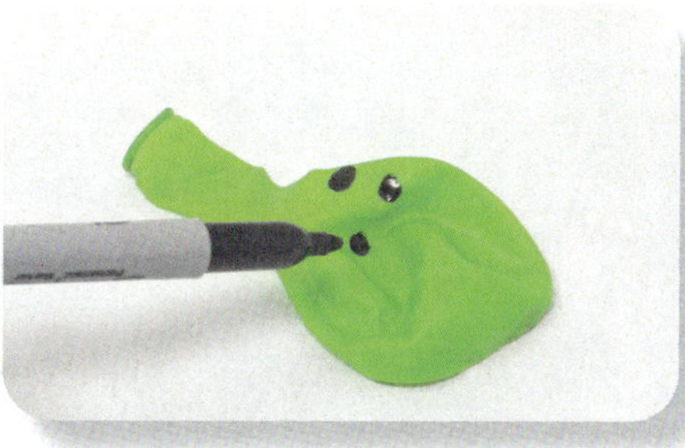

Step 2B

- Inflate the balloon until it is halfway filled with air.
- Tie a knot in the end of the balloon.
- Remember to ask an adult for help if you need it!

Step 3B

- Now, **observe** the dots all over the balloon. Are they still all the same size? Which dots became bigger?
- Where on the balloon are the latex **molecules** stretched out the most? Where are they stretched out the least? (Hint: Look at the size of the dots on the balloon. Where did the dot size change the most?)

Step 4B

- Dip 3 skewers into the soapy water. This will help the skewers to go through the balloon.The whole skewer should be covered with the soapy water.

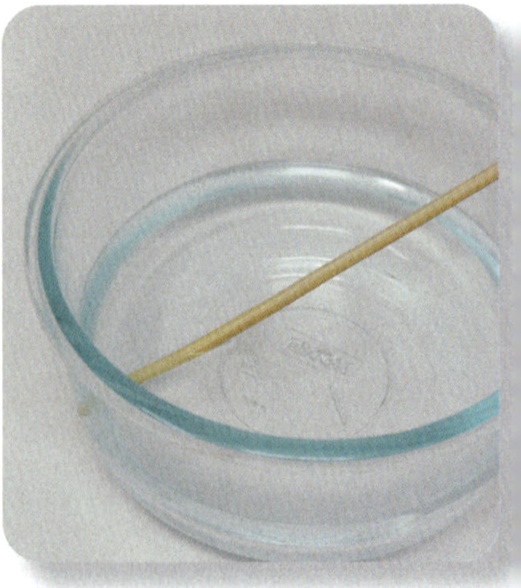

Step 5B

- Use the observations that you made earlier about the dots on the balloon to decide the best spot to pierce the balloon with the skewer. Can you put 3 skewers through the balloon without popping it?

Write down what you see

In your science journal, write your results!

- What was your hypothesis? (Look on page 12 of this book!)
- Was your hypothesis correct?
- Where did you pierce the balloon?
- How did you do it?
- Did the balloon pop?
- What happened to the dots when you inflated the balloon?

Why did this work?

- The secret to this experiment is *where* you pierced it with a skewer. Remember looking for the part of the balloon where the rubber was thicker? That is where the rubber **molecules** are under the least amount of stress.

- If you see the rubber that makes up a balloon on a **microscopic** level, you would see many long strands, or chains, of molecules. These long strands of molecules are called polymers. It is the **elasticity** of these polymer chains that causes rubber to stretch. The strands of **polymer** chains are stretched when you blow up the balloon.

SCIENCE

- Remember how the dots on the balloon changed size after you blew up the balloon? You probably noticed that the dots on either end of the balloon only got a little bigger. The dots in the middle of the balloon really stretched out and got bigger.
- You're a great scientist! You observed the area of least stress: the ends of the balloon! The points where you pierced the first time are where the polymer molecules were only stretched out a little. The long strands of molecules stretched around the skewer, and the air inside the balloon didn't come out.
- When you removed the skewer, you felt the air slowly escaping through the holes where the polymer strands were pushed apart. Eventually all the air came out and the balloon became deflated, but it never popped.

GLOSSARY

Elasticity: The ability of an object to return to its original shape after being stretched out.

To examine: To inspect something very closely.

Hypothesis: An educated guess; what we think will happen and why.

To inflate: To fill with air or gas.

Microscopic: So small you need a microscope to see it.

Molecule: A group of atoms that are connected together.

To observe: To watch something carefully.

Polymer: Large molecules that are made of small, repeating molecules.

Procedure: The steps to complete the experiment.

Question: What we want to know.

NEW IDEAS

How could we change this experiment?

- Could we replace the skewers with something else, like needles?

- What else could we use to make the skewers slide through the balloons easily?

- Could we pierce the balloon with many skewers to see when it pops?

Can you think of any ideas to improve the experiment?

- Make all the balloons the same size.

JOURNAL

Made in United States
Orlando, FL
07 November 2026

90337531R00017